ENGINEERING CHALLENGES

BUILDING FIGHTER JETS

by Marne Ventura

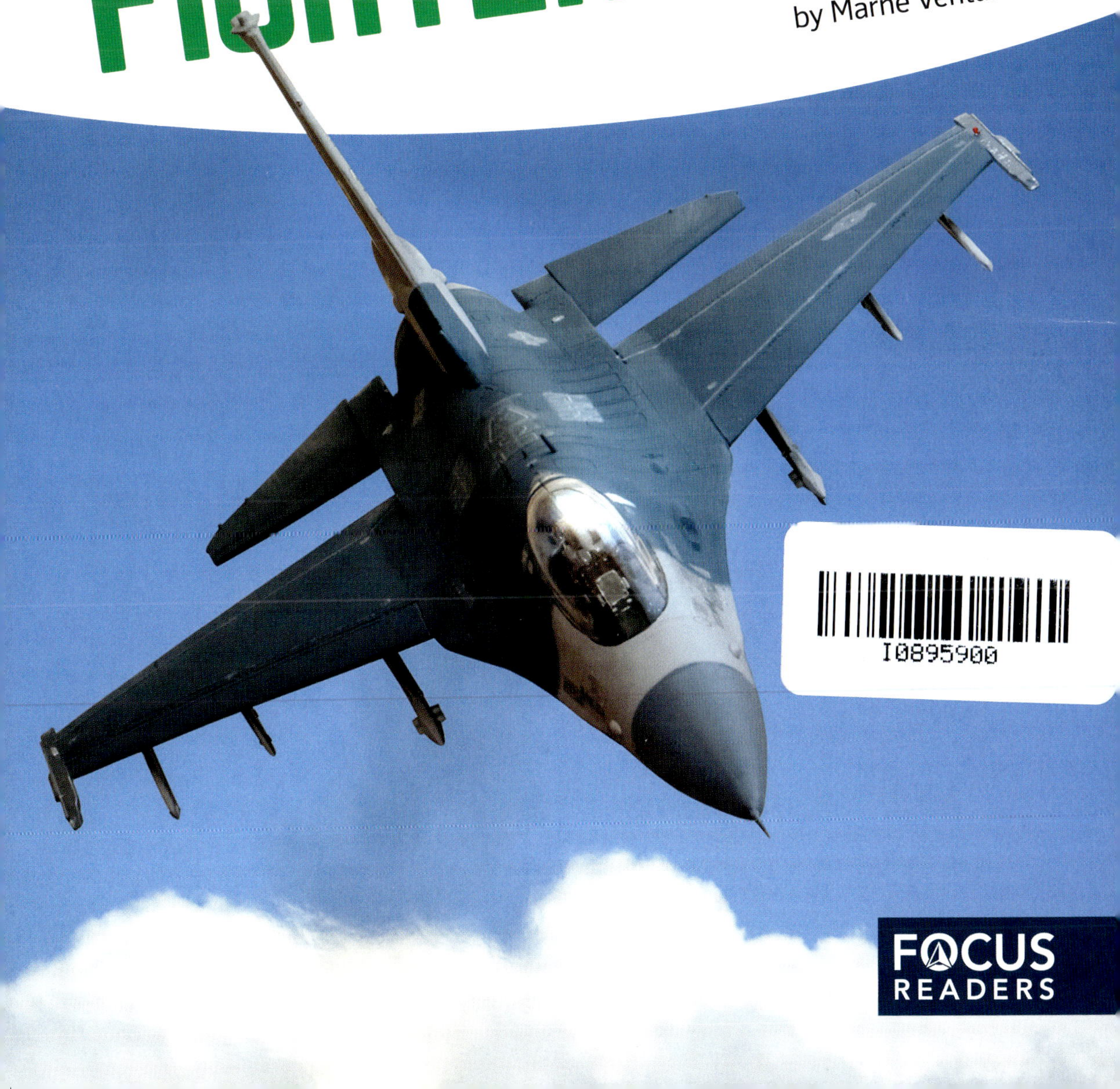

FOCUS READERS

FOCUS READERS

WWW.FOCUSREADERS.COM

Focus Readers is distributed by North Star Editions:
sales@northstareditions.com | 888-417-0195

Produced for Focus Readers by Red Line Editorial.

Content Consultant: James Flaten, PhD, Associate Director of NASA's Minnesota Space Grant Consortium and Contract Associate Professor in the Aerospace Engineering and Mechanics Department at the University of Minnesota – Twin Cities

Photographs ©: Andrew_Howe/iStockphoto, cover, 1; jax10289/iStockphoto, 4–5; AP Images, 7; guvendemir/iStockphoto, 9, 15; powerline1/iStockphoto, 10–11; CT757fan/iStockphoto, 13; Thomas D Dittmer/Shutterstock Images, 16; viledevil/iStockphoto, 18–19, 21; zoltanpocza/iStockphoto, 22–23; Red Line Editorial, 24–25, 29; Professor25/iStockphoto, 27

ISBN
978-1-63517-253-9 (hardcover)
978-1-63517-318-5 (paperback)
978-1-63517-448-9 (ebook pdf)
978-1-63517-383-3 (hosted ebook)

Library of Congress Control Number: 2017935921

Printed in the United States of America
Mankato, MN
June, 2017

ABOUT THE AUTHOR

Marne Ventura is the author of many books for kids. She enjoys writing about science, technology, arts and crafts, and the lives of creative people. A former elementary school teacher, Marne holds a master's degree in education from the University of California. She and her husband live on the central coast of California.

TABLE OF CONTENTS

CHAPTER 1

STREAKING ACROSS THE SKY

Ethan moved to the front of the bleachers and looked to his right. First he heard a rumble. Then he saw it. Nine fighter jets were flying in formation. They streaked over the crowd at the air show. Ethan cheered. The jets' power and speed was unlike anything he had seen before.

Fighter jets fly in formation during an air show.

Fighter jets are used by militaries. They are small, fast, jet-powered planes. A typical jet carries one or two people. Fighter jets also carry weapons, such as bombs or missiles. Different jets have different purposes. Some fight other aircraft in the air. Some are designed to attack targets on the ground. Some accompany other aircraft carrying supplies or people. Fighter jets defend these aircraft in case of attack.

Armies used propeller airplanes during World War I (1914–1918). Pilots often used handguns to fight enemies. Airplanes became more advanced during World War II (1939–1945). In 1944, the

A P-80 Shooting Star lands in 1945.

German military fought with Me 262s. They were the first fighter planes with jet engines. Fighter jets were much faster than propeller airplanes.

The US Air Force had to catch up. Engineers at Lockheed Corporation in California made the P-80 Shooting Star.

They designed and built the plane in only 143 days. The P-80 Shooting Star was still being tested in August 1945, about a month before World War II ended. But the Shooting Star was used in the Korean War (1950–1953).

In total, 1,715 Shooting Stars were built. They flew approximately 500 miles per hour (805 km/h) and carried six machine guns. They also carried either 10 rockets or a large bomb.

Fighter jets continue to get better with new technology. Today's fighter jets are fast and strong. But they are also lightweight and **maneuverable**. Some even have the ability to refuel in the air.

A fighter jet prepares to refuel in midair.

New technology helps pilots **navigate** and communicate. It also helps them find or avoid enemies during the night or day, in any weather. New designs and materials also make **stealth** jets possible.

CHAPTER 2

JET FIGHTER DESIGN

Most fighter jets have the same basic design. Wings lift the plane off the ground, allowing it to fly. Vertical and horizontal stabilizers form the tail of the plane. Stabilizers help control the plane's direction of flight. Movable parts on the wings and tail control the plane's altitude. Jet engines move the plane forward.

This Russian fighter jet has two vertical tail fins.

Additional parts support the plane while on the ground, taking off, or landing. The body, or fuselage, holds the pilot. The cockpit is where the pilot sits. This area contains the controls for the jet. Meanwhile, weapons are located in different parts of the plane, such as hanging from the wings or fuselage.

An aircraft is heavier than air. This means the force of **gravity** keeps it on the ground. So, planes must overcome the force of gravity through thrust and lift.

Jet engines create thrust. Air is sucked into the front of the engine with a large fan. Once inside, the air is **compressed** and mixed with jet fuel. Sparks ignite the

Jet engines propel this plane forward.

fuel. This causes burning gas to shoot out the back of the engine. Thrust pushes the jet forward so lift from the wings can overcome the weight of the jet. Thrust also overcomes **drag** created by **friction** with the air.

Jets are made of strong but light materials. This means there is less weight to overcome. Aluminum, steel, titanium, and fiberglass are all common materials.

Wings create lift. Their shape makes air flow faster over the top of the wing than it flows under the bottom. The air pressure on top of the wing is less than the air pressure under the wing. This pressure difference pushes the jet up.

Pilots need to make sharp turns at high speeds without losing control. Flaps, rudders, and other movable parts on the wings and tail change the way air moves around the aircraft. Pilots move these parts to make the plane change direction.

An F-15E Strike Eagle comes in to land.

Modern jets have computer systems that display information on the cockpit windshield. This allows pilots to keep their heads up at all times rather than having to look down at a control panel.

The B-2 bomber is known for its stealth design and construction.

Fighter jets use radar to locate targets. Radar works by using radio waves. The waves bounce off of objects. Stealth jets are able to avoid being detected by radar. They are designed with angled surfaces and covered with a special paint. These features prevent radio waves from bouncing back toward the radar source.

ENGINEERING DESIGN PROCESS

Engineers make sure the fighter jets they design are safe and effective. When they are satisfied with their design, the fighter jet is built.

ASK: What features are needed for a fighter jet? Which shape works best? Which materials are strongest and lightest?

IMAGINE: Brainstorm ideas for fighter jet designs. Make a list of important features.

PLAN: Draw a diagram of a fighter jet. Make a list of materials needed. Write down the steps to build the jet.

CREATE: Follow the plan and build the fighter jet. Make a test flight. Does it perform as expected?

IMPROVE: How could the design change to make it better? Try it again!

CHAPTER 3

BUILDING A FIGHTER JET

Most fighter jets take off and land by rolling down a long, flat runway. But runways aren't always available. So in the 1960s, British engineers designed and built the Harrier fighter jet.

The jet can take off by flying straight up. It lands by coming straight down. This allows it to operate from small spaces.

A Harrier jet performs in an air show.

The Harrier can take off and land from dirt roads or on short runways, such as a ship. The plane can also hover like a helicopter. It can fly sideways, backward, and even turn in place in midair.

To make these jets work, engineers added engine exhaust nozzles to different places on the fuselage. These nozzles can swivel to change the direction of the thrust from the engine. To move the jet straight up, the exhaust nozzles point downward. Gas shoots out toward the ground, which pushes the jet up. Changing the direction of different nozzles lets the pilot move in different directions.

The AV-8B Harrier II Plus has swept wings, which angle backward. With less drag, the plane can fly faster.

Engineers also used a special shape for Harrier fighter jets. The extra-broad wings are on top of the fuselage. They angle downward and have round edges. This shape helps the plane fly straight up or down. Movable flaps on the tail and wings also help steer a Harrier jet.

THE SUPER HORNET

The Super Hornet F/A-18 is the first US plane designed with the ability to be both a fighter (F) and attack (A) jet. It is built for air-to-air combat with other jets. It is also built to fire weapons from the air toward targets on the ground or at sea.

The F/A-18 has a strong airframe and landing gear. This allows it to take off and land on an **aircraft carrier**. Its wings are trapezoid-shaped and swept back. This gives the plane better performance and helps the pilots see better. In addition, the Super Hornet has a heads-up display and a high-thrust engine. Meanwhile, the Super Hornet has attack features, including missiles and a cannon. The jet can also fly at **supersonic** speed in any weather.

The first squadron of F/A-18 Super Hornets was sent into combat in 2002.

CHAPTER 4

FIGHTER JET CHALLENGE

Now it is your turn to build an airplane. Use the information you've learned about **aerodynamics** to design and build a paper glider.

Materials:

For this challenge, you'll need scissors and paper clips. You'll also need paper of different thicknesses or weights.

Paper gliders have the same general shape as real airplanes.

Procedure:

A basic paper airplane can be made in a few simple steps. Use these directions to build two paper airplanes. Start one by folding the paper the long way. Start the other by folding the paper the short way.

1. Fold the paper in half. This creates a crease down the middle. Open the paper to lie flat again.
2. On one end of the paper, fold both corners in to the center crease. This creates a point.
3. Make the point even sharper by folding both angled edges in to the center crease again.

Make sure your folds are accurate and firm.

4. Close the paper along the centerline crease. Fold each side down to create wings. The edge of the part you are folding down should be parallel with the bottom of the fuselage.

5. Fold the wings partly back up so they are horizontal, with the fuselage vertical beneath them.
6. Launch the airplane!

Improve It!

- Which shape flew faster and farther? Whichever one worked better, try rebuilding it to be even better.
- Think about the shape of the nose and wings. Are there different ways you could fold the plane to make these parts better? You can look online for more ideas and designs.
- Try cutting flaps or notches at different places in the wings, especially along the trailing edges.

The design of your paper glider affects how well it can fly.

- Adjust the weight and balance of the glider. Try adding paper clips to different places. Which locations result in the best overall performance of the glider?
- Try using lighter or heavier paper. Is one better than the other?

FOCUS ON

BUILDING FIGHTER JETS

Write your answers on a separate piece of paper.

1. Write a letter to a friend describing the main idea of Chapter 3.
2. Do you think new technology will be useful to fighter jet designers? Why or why not?
3. When was the first fighter plane with a jet engine built?
   - **A.** during World War I
   - **B.** during World War II
   - **C.** during the Korean War
4. How is a computer system that displays information on the cockpit window helpful to the pilot?
   - **A.** The pilot can fly in different directions in any weather.
   - **B.** The pilot can speed up or slow down quickly and easily.
   - **C.** The pilot can keep his or her head up and still see the display.

Answer key on page 32.

GLOSSARY

aerodynamics
The ability of an object to move through the air.

aircraft carrier
A warship on which aircraft can take off and land.

compressed
Pressed or squeezed together.

drag
A force that resists motion through a gas, like air, or through a fluid, like water.

friction
The resistance of one surface against another.

gravity
The force due to the mass of the earth that pulls objects downward.

maneuverable
Able to change direction easily and quickly.

navigate
To find one's way while traveling.

stealth
Designed to avoid being noticed.

supersonic
Moving faster than the speed of sound.

TO LEARN MORE

BOOKS

Collins, John M. *The New World Champion Paper Airplane Book.* Berkeley, CA: Ten Speed Press, 2013.

Perritano, John, and James Spears. *National Geographic Kids Everything Battles: Arm Yourself with Fierce Photos and Fascinating Facts.* Washington, DC: National Geographic Children's Books, 2013.

Portman, Michael. *Fighter Jets.* New York: Gareth Stevens Publishing, 2013.

NOTE TO EDUCATORS

Visit **www.focusreaders.com** to find lesson plans, activities, links, and other resources related to this title.

INDEX

Answer Key: 1. Answers will vary; **2.** Answers will vary; **3.** B; **4.** C